Mule Train Packing in Eastern Washington in the 1860's

The Century Magazine
Vol. XLI, No. 5 March 1891

Journal of Mule Train Packing in Eastern Washington in the 1860's

James W. Watt

Introduction by Lawrence Dodd

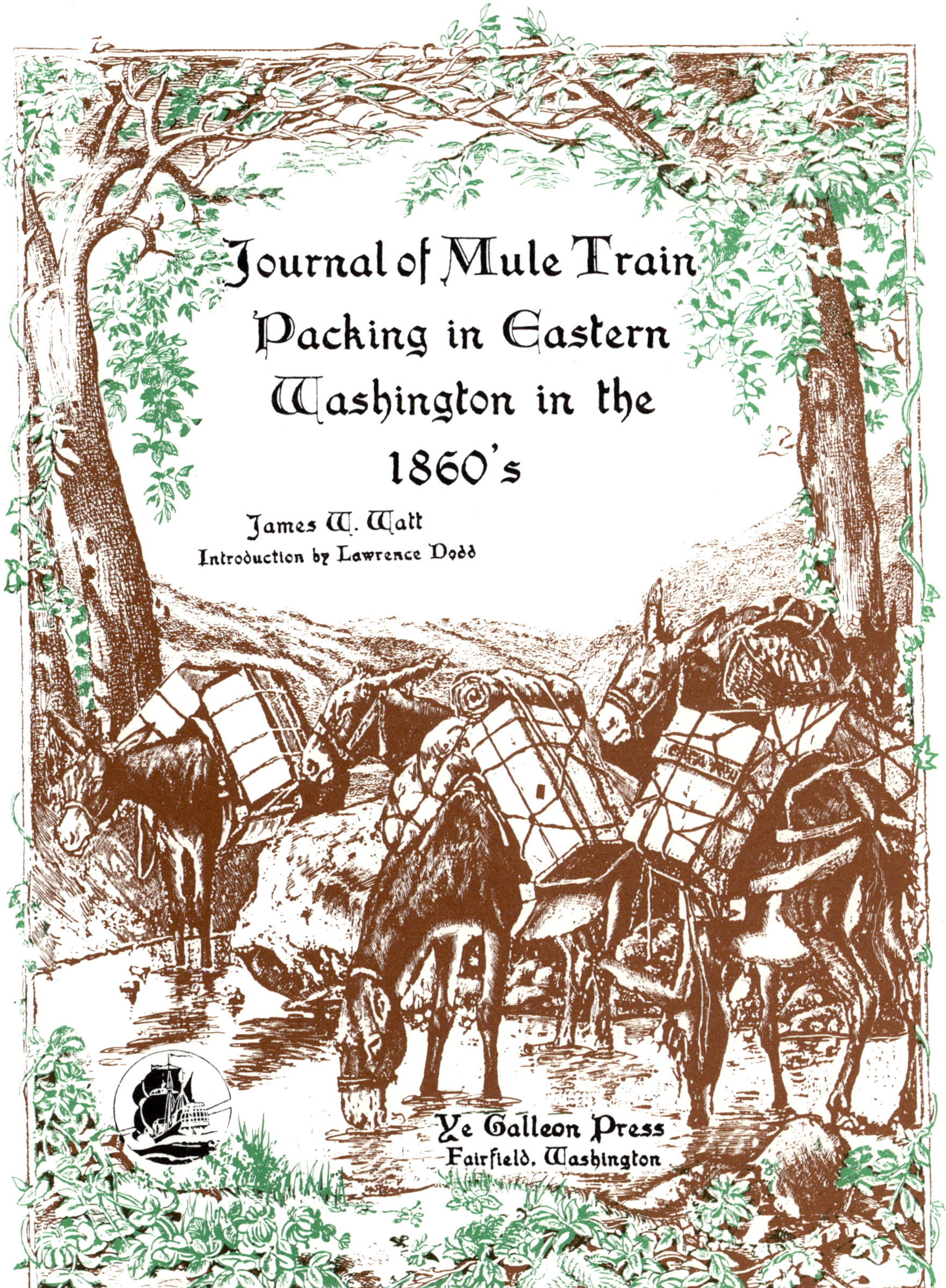

Ye Galleon Press
Fairfield, Washington

The James Watt journal was originally printed in *The Washington Historical Quarterly*, in three issues as follows:

Vol. XIX, No. 3, page 213 et al July, 1928
Vol. XIX, No. 4, page 285 et al October, 1928
Vol. XX, No. 1, page 36 et al January, 1929 **General**

Library of Congress Cataloging in Publication Data

Watt, James W., 1843-1944.
Mule train packing in eastern Washington in the 1860's.

Originally published in the Washington Historical Quarterly, 1928-1929.
Watt, James W., 1843-1944. 2. Pioneers—Washington (State)—Biography. 3. Muleteer—Washington (State)— Biography. 4. Washington (State)—History—To 1889. 5. Frontier and pioneer life—Washington (State). 6. Packing for shipment. 7. Washington (State)—Biography. I. Title.
F891.W47A35 979.7'03'0924 [B] 78-16012
ISBN 0-87770-205-5

Printed in U.S.A.
Ye Galleon Press
Fairfield, Washington
99012

Mule Train Packing

INTRODUCTION

The discovery of gold in the west was a time of great restless activity. Thousands of people streamed into gold bearing areas hoping to find a bonanza. As the population increased so did the area of exploration and with it the growth of mining camps and towns. The turbulent population pushed north into upper California and lower Oregon, east into Nevada, northeast into Idaho, eastern Oregon and Washington and north into British Columbia. Twelve years after the 1849 gold rush the Nez Perce Country was becoming the new bonanza. One of the thousands who hoped to capitalize upon this new discovery was James W. Watt, an 18 year old lad whose experiences in the Pacific Northwest gold rush have been preserved through the efforts of William S. Lewis (1876-1941.)

The James Watt narrative gives us an excellent insight into the character of the gold mining excitement during the 1860's. His story of experiences captures and almost forgotten era where men and animals shaped the face of the frontier. Along with describing the activities of miners and the camps and towns that served their needs, Mr. Watt explains the essential role played by the pack train.

Packing and the men who participated in this commercial venture were the product of a restless frontier. The first major gold discoveries were located in areas with easy access by water. Steamboats, using the seas and rivers, could easily deliver needed supplies to these early camps. But, as these gold bearing areas became exhausted and over populated, the miners migrated to remote locations. These new areas lacked navigable waters, had no roads and the trails were almost nonexistent. Consequently, new methods of moving supplies had to be employed and out of necessity the pack animal became the most essential element in the success of these early mining camps. This means of moving goods was certainly not a new idea in America. Americans had employed the services of pack animals for many years but not until the American gold rush packer encountered the Spanish did he learn the true art of packing. Utilizing Spanish techniques and equipment this new breed of packers began to establish a transportation system whose versatility became so efficient that they could meet the ever changing needs and conditions created by the mining enterprise. So, before the construction of adequate road systems the pack animal was the life blood of the mining areas. They were the connecting link between comfort and privation, plenty and starvation.

James W. Watt

The skills of the packer and his mule are aptly explained by S.J. Barrows in an 1875 article *THE NORTHWESTERN MULE AND HIS DRIVER*.—"Mule packing is a fine art. With a well-trained mule and a well-trained packer, there is nothing imaginable, from a bag of oats to a load of crockery, that cannot be securely fastened on the mule's back. Select the worst article you can think of to test the packer's skill, and it an incredibly short space of time he will pack it as though he had been perfectly used to packing that very thing all the days of his life. When the packer has finished, the animal may jump, back, kick, rear, or roll, to rid himself of his burden, but with no more success than Christian had before he reached the cross. And yet you cannot find a knot in the whole complexity of rope and bundle."

The packer would undertake the transportation of almost any article needed. Along with flour, gum boots, bacon, whiskey, and clothing a variety of unusual items were transported into the camps. It is recorded that one packer, utilizing four mules, transported a piano from Wallula to a sporting house in the Coeur d'Alene mining camps. The cargo arrived in excellent condition and the packer received $1,000.00 for his services. Another difficult cargo was a coil of 2" wire rope which was loaded at Umatilla and transported to a Montana ferry. The weight and length is not recorded but because the rope could not be cut it required three mules tied in tandem to carry the cargo. In the November 9, 1868 issue of *THE CITY JOURNAL* (Canyon City, Oregon) an article appeared stating "Messrs. Frank Louden (Louden) and (James) Madigan, with their pack train of two hundred animals,...have delivered at Harney,... three hundred thousand pounds of grain inside of five weeks." Mr. Watt tells about a 667 pound casting being carried on the back of one mule, while another source relates that "the heaviest weight a pack master confessed he had ever imposed on a mule was eight hundred pounds, a piece of machinery which could not be divided." Packers considered quicksilver one of the hardest items to pack and transport. This very dense commodity was normally shipped in iron flasks. Each flask of quicksilver weighed about 90 pounds and it was customary to pack two flasks per side. Special packing boards were used and constant care had to be taken to eliminate possible injury to the pack animal. So, supplies of all shapes and sizes found their way onto the backs of the pack animal. At first the packers were occupied in moving food, clothing and mining equipment, but as the wealth of the miners increased so did the desires for luxuries. Now, along with the necessities a weird assortment of goods were moved over the mountain trails and into the isolated mining camps.

During the 1860's Walla Walla, Umatilla, Wallula, Lewiston and many other supply points experienced a tremendous volume of pack animals coming and going daily. During the peak seasons supplies were being transported out of the supply towns as fast as they were coming in. Clues of this early activity can be found in the newspapers of the day. The April 7, 1865 issue of the *WALLA WALLA STATESMAN* notes that 262 pack animals were loaded and ready to depart the river port town of Wallula for various mining camps. Three years later the same newspaper reported that 23 separate pack-masters had or were loading a total of 1,000 animals. In the *WASHINGTON TERRITORIAL STATUTES, 1864-1867,* we find, in a memorial relative to the Mullan Road, the statement that "From Jan. 1 to No. 15, 1866...6,000 mules have left the Columbia River and Walla Walla loaded with freight for Montana." This large volume of pack animals picking up supplies was not exclusive to Walla Walla, Umatilla and Wallula. All the supply towns were experiencing the same traffic. Mr. Watt's narrative will give you a good insight into these years in which the pack trains held a monopoly on the transportation of goods into the remote mining regions.

One of Mr. Watt's contemporaries was Ben Burgunder. On July 14, 1907 Mr. Burgunder delivered an address before the Inland Empire Pioneer Society (Walla Walla) giving the following tribute to the pioneer packers:

> "There is one class of the early pioneers of the Pacific Coast whom the writers of history have failed to mention. The early packers are, in my opinion, among the foremost of the pioneers of the Pacific Coast, and they did as much, if not more, than any other one class of men toward making the great Northwest what it is today."
>
> "They were as honest and as brave a set of men as ever lived, and without them the Inland Empire could never have made the progress that it has. They were all men who did not know the meaning of the word fear, and they risked their lives and property to serve the public. Without them it would have been impossible for the merchants to carry on their business."

James W. Watt

"Indeed, the pioneer miner, farmer or anyone else could not have lived had it not been for the help of the brave and hardy packer. As there were no wagon roads in those days all freight had to be transported by pack horses, and the packers had a hard and dangerous time furnishing the people with meat. They had to be out in all kinds of weather and endure a great many hardships. Often it was even necessary for them to fight their way through the Indian country. Those familiar with the country in the days when it was first settled know how it was the packers and traders that were always foremost in blazing the trails and preparing the way for civilization. To such men—I pay a loving and grateful tribute."

Mr. Watt's narrative is a valuable and important contribution to the annals of the gold rush era. His experiences have given us another tool by which we can better understand the development of our country. All who read and use James Watt's narrative should remember that if it had not been for the two years of interviewing by William S. Lewis and the cooperation of James W. Watt, this account would never have been written. Now , through the press of Glen Adams, this important material has again appeared in print. All these men deserve our thanks.

Lawrence L. Dodd

Mule Train Packing

I CAME TO WASHINGTON territory for the first time in 1860. In the fall of 1859 and the spring of 1860 I brought up a herd of cattle over the old "Emigrant" (Barlow) road to Tie Valley, just south of The Dalles, Oregon. I was back in Yamhill County again when we got news of the discovery of the Pierce City gold mines. Many of the men in our neighborhood had been in the California mines and most of these, including several of my relatives, started at once for the new Eldorado. In July, 1861, I came up to the mines in company with a Scotchman named John McClaine, who had been working for my cousin, D. M. Jesse. Mr. Jesse had gone in to the mines with the first rush in the early spring of 1861 along with my brother, Alex Watt, and he was then engaged in conducting a general merchandise business at Oro Fino under the firm name of D. M. Jesse & Co.

At Walla Walla we joined the pack train of D. M. Jesse & Co., and went into the mines with it, via the site of Lewiston, Idaho. There was nothing there then but a camping ground. We reached the mining town, Oro Fino, about July 20, 1861. I remained there until October, 1861. The Pierce City or Oro Fino placer mines were situated on the North fork of the Clearwater River, along Oro Fino Creek and its tributaries, and the "diggings" extended along Oro Fino Creek, and up the various gulches, for a distance of some twelve miles above Pierce City and for some ten miles, or more, down the creek below that town.

By the summer of 1861, when I went in, there had already been a large amount of work done on the trail from the mouth of the Clearwater (later the site of Lewiston) to Oro Fino and Pierce City, evidencing a determination on the part of the people to put it in the best possible condition. This work was done under the direction of a Mr. Athey from Oregon City, Oregon. After great effort some twelve or fifteen wagons were finally dragged up to Oro Fino. Mr. Mulkey from Washington County, Oregon, drove the first of those teams into Oro Fino. However, it was a mighty poor road for wagons, and much more time and money, than was then available, would have to be expended on the trail before it was feasible to haul supplies in over it by wagons. Consequently everything in the mines had to be transported into those remote mining regions on the backs of mules. All goods destined for Oro Fino, as well as for Florence in later years, were transported by pack trains in this manner. There were no practicable wagon roads then leading to these mines.

The miners estimated that there were all of twelve thousand people in the Pierce City or Oro Fino Mining camp by the end of July, 1861. Oro Fino itself was already quite a town when I arrived. The population in and about the town numbered around twelve hundred; Pierce City was as large or larger. About four hundred houses and tents had been erected in Oro Fino by August, 1861, and the place was improving fast. Most lines of business were already established. I had more or less contact with the business life of the town, and at one time or another I personally met most of the people then in business there.

I will try to describe the town as I knew it in 1861. There were some fifteen or sixteen different stores of various kinds; about eighteen or twenty saloons, six bakeries, four hotels, three meat markets, five laundries, two bankers, four doctors and two express offices,—Tracy & Co., and Mossman & Co. There were no lawyers as I remember, at least none attempting to practice, and no Chinamen as yet. As was customary in these mining towns, there were almost as many saloons as there were stores. Nearly everyone drank to some extent then. There were plenty of gamblers, a dance hall or two, and the usual women of a camp. I recall in particular two of the larger saloons where gambling was extensively carried on, and all sorts of devices were resorted to for the purpose of enticing men into these places. There was much drinking and gambling and many drunken rows. Shooting scrapes were of rather frequent occurrence.

The buildings were mostly frames covered with canvas, or log shacks. Lumber was very scarce and expensive, as it all had to be whipsawed by hand. Much of it sold as high as $1.25 per foot. Of the four general merchandise stores in Oro Fino in 1861, that of my cousin, D. M. Jesse & Co.'s, was one of the larger. Most of our business was done on Sunday, and on that day we used to take in lots of money—gold dust. Occasionally a miner was broke, and bought "on bedrock;" that is, he was given credit until he got down to pay dirt and could take the gold out of his claim. At first the miners, principally from Oregon and California were pretty honest, but later on we had a more tricky bunch to deal with, and merchants had to watch out for such things as brass filings in gold dust, and had to refuse credit to strangers.

It was customary for miners of these camps to make "their clean-up" on Saturday afternoons—that is clean out their sluice boxes and gather the gold dust and nuggets that had collected in the riffles. Some of this gold was so fine that it had to be caught with quicksilver. Sunday was the miner's day off and the busy day for all those in the mining towns. No store, shop, restaurant, saloon, dance hall, gambling place or other business establishment closed its doors on Sunday. On that day the regular town population was doubled and trebled by the influx of miners who trooped in from all the various gulches and diggings up and down the creek to get their mail, to purchase supplies, and to enjoy themselves in social gatherings with friends, and the diversions of the saloons, gambling tables, dance halls and other pastimes and dissipations offered by the town.

For nearly four months I remained in the camp. During much of this time I was engaged in delivering goods from D. M. Jesse & Co's. store at Oro Fino to the miners in neighboring diggings who came into town on Sundays and bought their supplies. My business thus took me into all the various gulches and diggings in the neighborhood from point about three miles above Oro Fino to a short distance below the town; the stores at Pierce City took care of the trade from there down. There was an abundance of provisions packed in the Oro Fino camp by the summer of 1861, and prices were very reasonable considering the high cost of freight. Flour for a time was sold for only 16 to 18 cents a pound, bacon was down to 35 or 40 cents, beef at 12 to 15 cents, sugar at about 30 cents and tea around $2.00 a pound. Everything else was in proportion.

During the summer and fall of 1861, when I was there, all of 2500 practical miners were actually at work on claims along Oro Fino creek and the gulches within a distance of twenty miles up and down Oro Fino Creek. There were hundreds—yes, thousands of other men employed at digging ditches, whipsawing lumber, clearing off claims, and prospecting about the country. The principal gulches in the camp were Rhodes Gulch, Canal Gulch and French Creek. My own personal knowledge and observation, while general, was particularly directed to Oro Fino Flat and Rhodes Gulch by my more frequent visits to these diggings.

DEAD MULE TRAIL, IDAHO—From a Sketch by an Army Officer.

Mule Train Packing

On Oro Fino Flat some poor claims only produced from $6.00 to $10.00 per day to the man, but on Rhodes Creek, which empties into Oro Fino Creek above Pierce City, the claims produced from $20.00 to $60.00 or more per day to the man. Some exceptional ground there produced even $75.00 to $100.00 per man per day; the average however was much lower.

I recently read a contemporary statement, concerning the production from these mines, made by C. C. Robbins of Portland, Oregon, in August 1861, and published in the *Oregonian* of August 26, 1861. This reads as follows:—

"The yield of the mines is as follows—Jarvis & Co., four men, $10. per day to the man; James & Co., five men, $10. per day to the man; McCarty & Co., four men $10. each; Vesay & Co., eight men, $7. to $8.; Hook & Co., six men, $10. to $12.; Jones & Co., four men, $10. to $12; Dunbar & Asar, $10. to $12; Shaffer & Co., fourteen men, $60.; Paine & Co., twenty men, $70.; Mortimer & Co., twenty-four men, $70 to $80.; Hatch & Co., five men, $16. to $20.; Thomas & Co., fourteen men, $16.; Boon & Co., eight men, $16.; California Co., nine men, $16.; Newland & Co., six men, $16.; Hickox & Co., five men, $16. to $20; Let 'Er Rip Co., eleven men, $16. to $20.; Hoyt & Co., eight men, $12.; Felton & Co., $16.; Sparks & Co., $15.; Rossi & Co., $15.; Rhodes & Co., eleven men 300 ounces per day to the company. On French Creek, Antoine Pillir, T. Lepoint, M. Guinon, John Lesot, Markum & Quirk were making each $10. to $12 per day."

I did business with many of these people, and I consider Mr. Robbins' statement a fairly accurate statement of the actual gold production per man in the Oro Fino camps at the time stated, August 1861. I was there at that time. It would be hard for me to make any independent estimate of the gold actually mined in the Oro Fino and Pierce City districts during the time I was there in camp. The average production per man in the months of July, August, September and October, 1861, when one considers the large production by companies and from exceptionally rich ground, at a most conservative estimate was certainly over $10.00 per day. Even at that very conservative figure the total production probably passed $25,000. or more, a day, in the four months I was there in the camp.

Most of the miners, who had claims and worked them, did pretty well, and the claims in the district that were opened, take them all, so far as I could see averaged at least $10. a day to the man. There were some big strikes say from $50. to $75. or more per day to the man. This was principally on

Rhodes Creek. All the 2500 or more practical miners then working claims in the district appeared to me to be doing well and were in a fair way to "make their pile," as the Californians used to say. During the latter part of the summer and early fall there was some little excitement and talk among us about discoveries on the south fork of the Clearwater. There were supposed to be at least two thousand men down there by that time and for some weeks we heard of good ground being struck about Elk City, but we didn't pay much attention to the South Fork country until the Florence diggings were discovered on the Salmon River later that fall.

As I have said, all of us in Pierce City and Oro Fino were pretty well satisfied and doing well when all of a sudden news of the rich Salmon River strikes came in. Everyone who could set out for the new diggings and immediately the Oro Fino camp was pretty nearly deserted.

The "Salmon River mines" was a name we applied to a considerable stretch of country including Miller's strike at Millersburg on Miller Creek, and the later strike at Florence—some 30 miles away. Later we generally referred to the whole district as the Florence district, or Florence mines, and I use the word Florence in that general sense.

Many of those who went couldn't get good ground at Florence and finally came back to their old claims at Oro Fino. We, at D. M. Jesse & Co's store, had advance news of the discovery from friends in the new diggings. This news was whispered around among a favored few and we went in on the first excitement, before the general rush which was about October 15, 1861. I set out for Florence with Colonel Althouse, an old school teacher from Yamhill County, Oregon, who was then employed as a clerk in my cousin's store at Oro Fino. He lived neighbor to D. M. Jesse whose home was two miles south of Lafayette, Yamhill County, Oregon. During the summer of 1861, Mr. Jesse had returned home to Oregon for a visit and hed left Col. Althouse in charge of the pack trains and freight.

On the first news of the Florence discovery two pack trains of supplies were sent out for the new camp by enterprising merchants. One from Lewistown, owned by Bledsoe and Creighton, consisting of 85 animals, and the other from Oro Fino, consisting of 35 animals, and owned by Jesse and Thompson. This latter train, under charge of Colonel Althouse, was the one I accompanied. The Florence Mines were 85 miles by trail from Oro Fino and it took us about six days to make the trip. The Bledsoe and Creighton pack train came up from Lewistown over the Lewiston trail (about 100 miles from Florence) and we intercepted it on the road and all went in to Florence together.

We arrived at Florence about October 25, 1861. There had been an early cold spell, and there was then about 15 inches of snow on the ground. Only 200 or 300 people were there when we arrived but more kept coming in rapidly and soon there were thousands. As a consequence of this great influx of people, many of whom brought little if any thing in the way of supplies, the prices of good rose tremendously. Flour was soon $100 to $125 for each hundred pounds, and other goods in proportion.

This Florence camp was generally regarded as one of the richest placer gold strikes ever made in this part of the country, although the gold dust itself was not as valuable as the finer gold obtained from some of the other camps. Florence gold was worth about $14. an ounce in trade. Pay dirt was struck about two feet deep, with two or three feet of loam and top soil on top; bed rock was reached at five to six feet.

My brother Alex Watt and his partner Billy Morris already had three good claims at Oro Fino; they abandoned these to go to Florence; even Bill Rhodes left his rich claims on Rhodes Gulch and joined the stampede. To show the richness of these new diggings at Florence, my wife's uncle, John Monroe, bought a claim on Summit Flat; this he worked with a rocker and but 3 or 4 men. He had a buckskin poke or purse that would hold about $2,000 in gold dust and he would have this filled every day with gold from the claim. Such production was about the average on Summit Flat or prairie while I was there.

Quite a few notorious characters came into the Florence camp that fall and there was considerable trouble. I'll tell you of an incident I witnessed. Early in the rush three parties located a claim on lower Baboon gulch, about a mile and a half below Florence. Two of these men were brothers, whose names I've forgotten, the other man was named Lyons. Like nearly everyone else in camp they were short of provisions so it was agreed between them that Lyons should take a horse and go down to Lewistown for supplies while the two brothers remained in camp to protect and work the claim. Lyons got back from Lewistown with the supplies about the time we arrived. Everyone was then getting ready for winter; there was already some twelve to fifteen inches of snow on the ground, and some enterprising individual had just driven in a band of cattle to be slaughtered for winter meat.

Well to get back to Lyons:—when he returned to the claim he found that the two brothers had "frozen him out." He stuck around and insisted on being taken in. They had a heated quarrel. Finally one of the brothers shot and killed Lyons while he had an axe in his hands clearing some dead timber on the claim. Colonel Althouse and I were in our camp eating our dinner when we heard the shot. We finished eating and then saw a crowd of miners gathering at the head of the gulch a few yards away. I went over there and saw the dead man, Lyons, lying on the blood stained snow. The sight made me sick. Lyons was the first man I ever saw that was killed in cold blood. Since then I've seen many similar sights and have to some extent become hardened to such things.

The assembled miners held a mass meeting right away and chose Bill (William) Mayfield, whose brother was a Baptist minister back in Oregon, as sheriff. They gave Bill $200. to arrest the murderers and everyone thought it would be a dangerous undertaking. Bill Mayfield was a man somewhat noted for his bravery; he just walked over to their cabin, which stood in the middle of the gulch, and hollered out to them that he wanted them to come out and "to give up." They came out at once and surrendered to Bill without resistance and stood trial. A hearing was had that night. The brothers claimed that Lyons had been shot in self defense, and that he had threatened one of them with the axe, so the miners agreed to let them off, and had Bill release his prisoners. William Mayfield was killed in Bannock City a few years later.

The next morning Colonel Althouse and I were out and at work by daylight and got an early start with our pack animals. We camped for the day early, over at the head of Slate creek. We were just making our camp when these two brothers came along the trail with their blankets and packs on their backs, headed out of the country. They must have had a premonition or warning to get out of Florence and to get out quick. Some time later a party of about twenty-five armed men came along and inquired of us for the two men. Later this same party of men came back by our camp. They weren't talkative, but one of them remarked that the two men "wouldn't fleece any more partners." We inferred they had lynched the men. This was my first encounter with the miners' law.

This murdered man, Lyons, had been a Mason, and after his death the Masons at Florence held a meeting and appointed a committee to look after Lyons' claim. Among the personal effects in Lyons' pockets were some letters giving the name and address of a niece living somewhere in California. The committee wrote to her that they had taken charge of Lyons' claim and that they were offered $16,000. for it, and inquired what his heirs wished to be done. The reply came back to sell the claim, and they did, sending the money back to Lyons' relatives in California. This Colonel Althouse, who was with me, was rather prominent among the Masons in the early Idaho mines.

Colonel Althouse and I continued on back to Oro Fino with the pack train. Early in the winter I left that camp and returned to Oregon. I spent the winter of 1861-62 in Yamhill County, Oregon. This was one of the worst winters ever experienced in this country. Between The Dalles and Walla Walla thousands of cattle perished, dying for lack of food and water. There was a very heavy snow that winter, and the break did not come until February. In Eastern Oregon the cattle would come down from the snow-covered hills seeking water, but the streams were frozen over. Hundreds died along these streams, and in the spring the dead bodies were washed away by the spring floods.

That winter many careless and improvident miners in the Florence and Elk City Mining district perished. One party, setting out from Lewistown for Portland was caught in a blizzard and froze to death on the sage brush flat at Butter Creek, near Umatilla. Navigation on the Upper Columbia had to be suspended, and for a time even mail and express was held up. The middle of January, 1861, my brother Alex Watt, carrying the express and some mail on his back, got out of the upper country and came down the river on the *Brother Jonathan* to Portland.

In the following year I began to pack goods from Walla Walla to the placer mining camp at Oro Fino, which was then in Washington Territory, but now the State of Idaho. After the Florence placers were struck in October, 1861, goods were transported thither from Walla Walla and Lewiston. Navigation was opened on the Snake River as far as Lewiston in 1862. The town of Lewiston grew rapidly. At first it was just a town of tents and shacks. I was first on the site of Lewiston in July, 1861, as I have said, but there was at that time no evidence of a town there then. It started as a camping place. Hill Beachy built the first hotel in Lewiston, the "Luna House," it was just a frame covered with canvas. The population reached around 600 to 700 in 1865-6.

"Mary Moody" on Pend Oreille in 1867. Pack Train boarding. Reproduced from the Oct. 1965, issue of Pacific Northwest Quarterly.

UMATILLA IN 1864.

From Umatilla & Morrow Counties by W.S. Shiach 1902.

Mule Train Packing

Lewiston was then the headquarters for a number of express lines. Dwight Bros. owned and ran Hill Beechey's old Lewiston stables, and H. Criter and T. J. Moore had the other livery stables. These stables were the headquarters for Davidson's Oro Fino and Pierce City express, Fairchild's Elk City express, and Ramey & Co's Florence and Warrans express, and Waldron & Co's Montana express.

In 1863 I began packing goods into the mining camps of the Boise Basin. Umatilla, on the Columbia River, was the chief outfitting point for the Boise Basin Mines, although considerable merchandise, particularly flour, was transported from Walla Walla to those mines. Umatilla, was the center for the heavy freight business into Eastern Oregon, Southern Idaho and northwestern Nevada. J. H. Koontz was postmaster there, in 1865-1866. J. H. King and Thomas & Stanfield ran the livery stables; and J. H. Hailey & Co., operated a stage line into Boise. The distance from Umatilla to Boise Basin was about 300 miles and the price received for carrying freight over this route ranged at first all the way from 40 cents to 80 cents a pound. There were many pack trains on this route. During the first year or two the Indians caused us considerable annoyance. I was on this route for more than two years. During this period freight was carried inland from Umatilla as far as Winnamuca, Nevada.

On these packing trips we seldom carried food for the mules that were employed in packing, but depended upon the grass produced by the regions through which we passed. This native forage usually kept the animals in fine condition. The average pack mule sold for $250. A very good mule would bring $400.00. The wages of the packers varied from $100 to $125 a month, in addition to board. The train master would frequently receive as much as $150 a month. Gold dust was then the only legal tender used; greenbacks were at a heavy discount, and in great disfavor. To attempt to pass them as legal tender at par was considered a shady trick. In fact some merchants even went so far as to blacklist and publish the names of persons using greenbacks to discharge obligations incurred at gold prices. Freight and wages were thus always paid in gold and as a result many of the pack trains carried quite an amount of gold dust and they were frequently held up and robbed by road agents.

On the Boise pack trail in early days some boys had a cayuse pack train. I made one trip with that train. It was a poor outfit. Mules were much easier to work with, and they would carry much heavier loads than the Indian

This old mule train drawing is from Harper's Weekly, October 12, 1878.

cayuses. Two years later on the Boise trail I found the spot where John Welsh, a packer, had been held up by road agents and murdered. The place was a mile and a half the other side of Grass Valley. In coming across the rye grass flat, Welch's pack train was held up. Welch had $1600 in gold dust, and a companion had $3,000. in greenbacks—then worth 40 cents on the dollar. When commanded to hand over his money, John said, "I'll see you again." "No you won't," replied one of the robbers, and he blew Welch's head off with a shot from a double barreled shot gun. This was, if I recall correctly, in 1864 and I believe that an account of the murder was published in a contemporary issue of the *Oregonian* at Portland, Oregon.

I was on the trails for ten years but was never attacked by Indians or road agents. The Boise Road was considered the most dangerous from attacks of both. The Bannock and Snake river Indians made frequent raids on the Boise trains, and I have often come along just after an Indian attack or a highway robbery. In this country it was customary to set out night riders to look after the animals. The Indians frequently attacked and killed the herders and ran off the stock; they made such raids all the way along the road from Powder River to the Boise Basin.

One time in 1864 I was camped above Burnt River on some high ground near Miller's station, a well known stopping point on the Boise trail. On the low ground near me there were some cayuse trains. During the night Indians stampeded 400 or 500 head of pack animals. We heard the animals coming up the gulch we were in driven by the yelling Indians. We got our own mules safely into camp just as the Indians came into sight on the high ridge above us.

The holdups on this route used to hold up whole pack trains, slashing the aparejoes and cutting up the rigging trying to find hidden gold. These aparejoes had a hole in the center, usually stuffed with wild hay for packing. It was customary for a packer to pull out the hay and hide his gold dust in one of the aparejoes, and the robbers would often cut up and ruin 40 or 50 aparejoes worth from $50 to $75 each in an effort to find hidden gold dust.

Later in 1864 I went into Elk City at the head water of the South Fork of the Clearwater River with a pack train of freight. Elk City was located between Elk and American Creeks of the Red River; the main branch of the Clearwater. It was a difficult trail from Oro Fino, over three mountain ridges, and the distance was figured at 120 miles. The population was around 500 or 600 people at that time. We arrived in town in time to partake in a celebration. Four Germans, who had successfully mined in the camp, and "made their stake," had just sold out their claims and were on a tear. They had chartered all the saloons and hotels in the town, and were offering free food and drink to everyone. They had received between $80,000 and $100,000 in cash for their placer ground, and celebrated their good fortune in true Germanic style. They insisted that everyone in the community join them in having a good time. Every thing, every where, was free. This celebration continued for several days after our arrival, when the happy Germans at length took the trail towards Lewiston, on their way back to the Fatherland, with the best wishes of everyone in Elk City.

I made one trip over the Mullan Road into Helena, Montana in the fall of 1865. This was right after the close of the Civil War. Helena was quite a city, even in those days; the population was around eight thousand people. It would be pretty difficult now-a-days for anyone to form any adequate conception of the activity and bustle in some of these early mining camps such as Florence, Helena, or the Boise Basin at the height of their gold booms. The few streets were jammed with traffic. Pack trains were arriving, unloading freight in front of the stores or departing; the stage coaches would arrive; little groups of new comers were constantly coming in while others were departing with loaded pack horses and outfits for mining and prospecting ventures. Everyone was in a great hurry. New buildings were going up all around and one heard a continual pounding of hammers. The hurdy-gurdys and saloon orchestras were playing. The promotor was loudly calling attention to his town lots or mining stocks. The saloons and dance halls were all going full blast and the whole noisy, hustling hilarious crowd gave an impression and created an effect of hustle and great business activity not equaled by the crowds on State Street in Chicago or on Broadway, New York.

The post office was established in Helena that year, 1865. Before that all mail had been brought in from Salt Lake City by the pony or stage expresses at a cost of 25¢ to 50¢ a letter. My cousin, D.M. Jesse, then had a store there. I saw a number of the notorious characters about the town. I recall one man named Bummister—a German. I had heard of him before in the Kootenai country in British Columbia. He had killed an Irishman there, and hastily quitting that section he came to Helena. There he got into a mining ditch company, and was soon reputed to be worth considerable money. He finally wormed himself into a position of some prominence in the community, and even got to be a guard or minor officer in the vigilante organization. One day a man named Kam Kerby, who had known him in the Kootenays, encountered Bill sitting on a beer keg in front of a saloon and he wanted to borrow a dollar or two from him. When Bummister refused him, Kerby said "You —, —, —, why my kindness saved your neck, coming out of the Kootenays." If this was true Bummister returned the favor in a singular way. Kam had a brother, Bill Kerby, a noted desperado who was then absent bringing a couple of sporting women into camp, and he was therefore not in touch with the recent vigilante activities. The vigilantes were just waiting for Bill Kerby to come back to Helena so they could hang him. Kam, learning of this in some way secured a horse, and going out, intercepted his brother Bill on the trail and had him turn back and leave the country. When Kam Kerby got back to Helena, this Bummister informed the vigilantes against him as an outlaw character and they swung him up on the same old pine tree in Dry Gulch where 11 or 12 other men had been hanged by the vigilantes.

One day while I was unloading freight in front of D.M. Jesse's store at Helena, I heard a personal altercation between this Bummister and another man. Two men, strangers to me, but evidently old acquaintances of each other had met and were talking together on the street some five or six steps away from me. One of these men made some remark about the war that didn't please Bummister, who chanced to be passing by at the time and heard. Bummister stopped and called the man a hard name. The stranger turned to him and said that if he repeated the epithet he'd knock him down. "Who, the Hell are you, anyway?" the stranger added. Bummister pompously answered, "William Bummister." "Well," said the stranger, with contemptuous emphasis, "I don't think much of either you or your name," and it looked as though Bummister had a fight on his hands.

This all took place on the side walk and the loud talking attracted quite a few people. Bummister seeing he was in for trouble gave the secret sign of the vigilantes, and quite a few men connected with the organization went to his support. This man, who was a packer, told them in a frank convincing way just how the trouble with Bummister had occurred. One of the vigilantes walked over to the packer's side and soon most of the crowd were with him. One of the vigilantes who seemed to be a leader then consulted with two or three others in the crowd, and then turning to Bill Bummister told him that

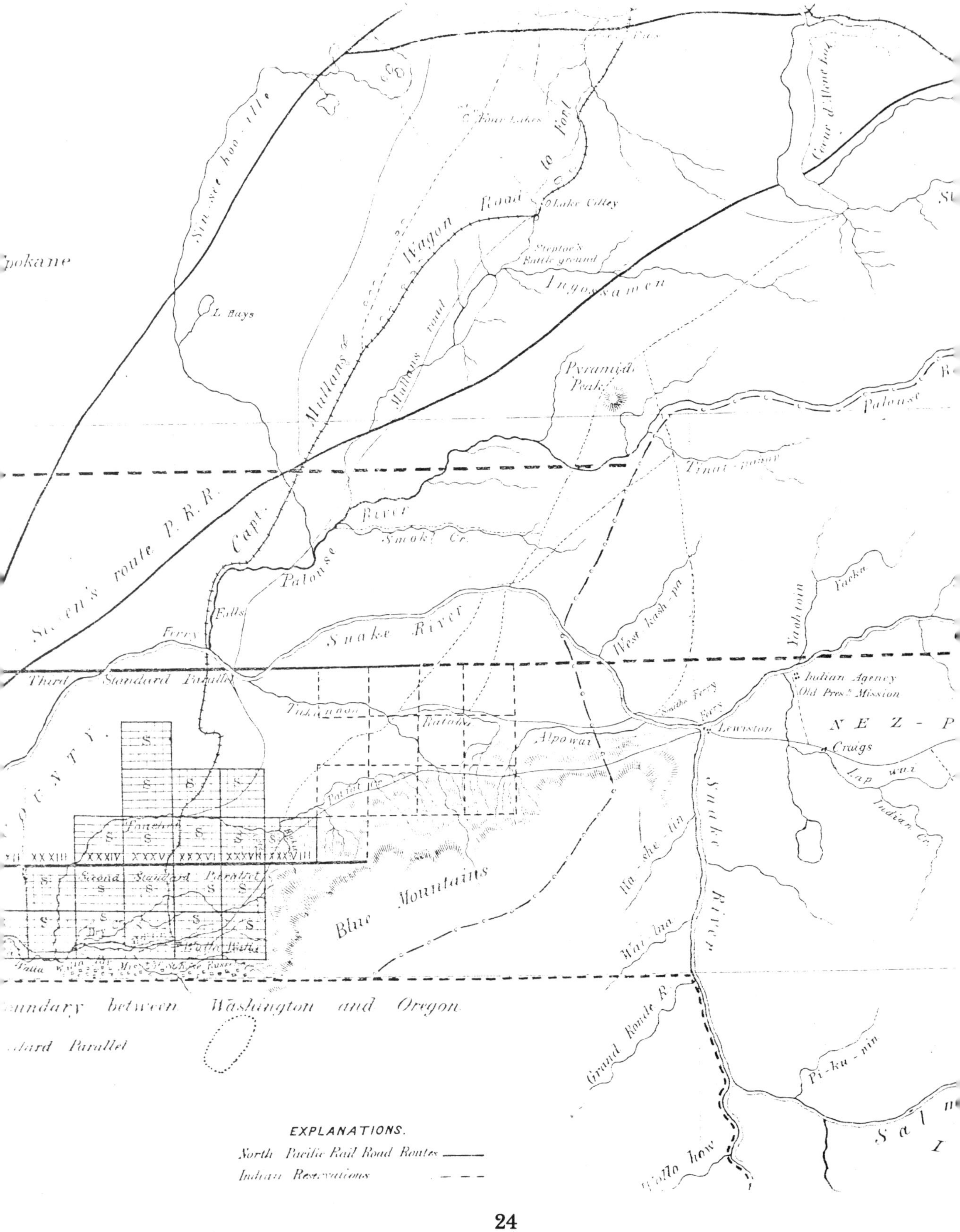

EXPLANATIONS.

North Pacific Rail Road Routes ————

Indian Reservations – – – –

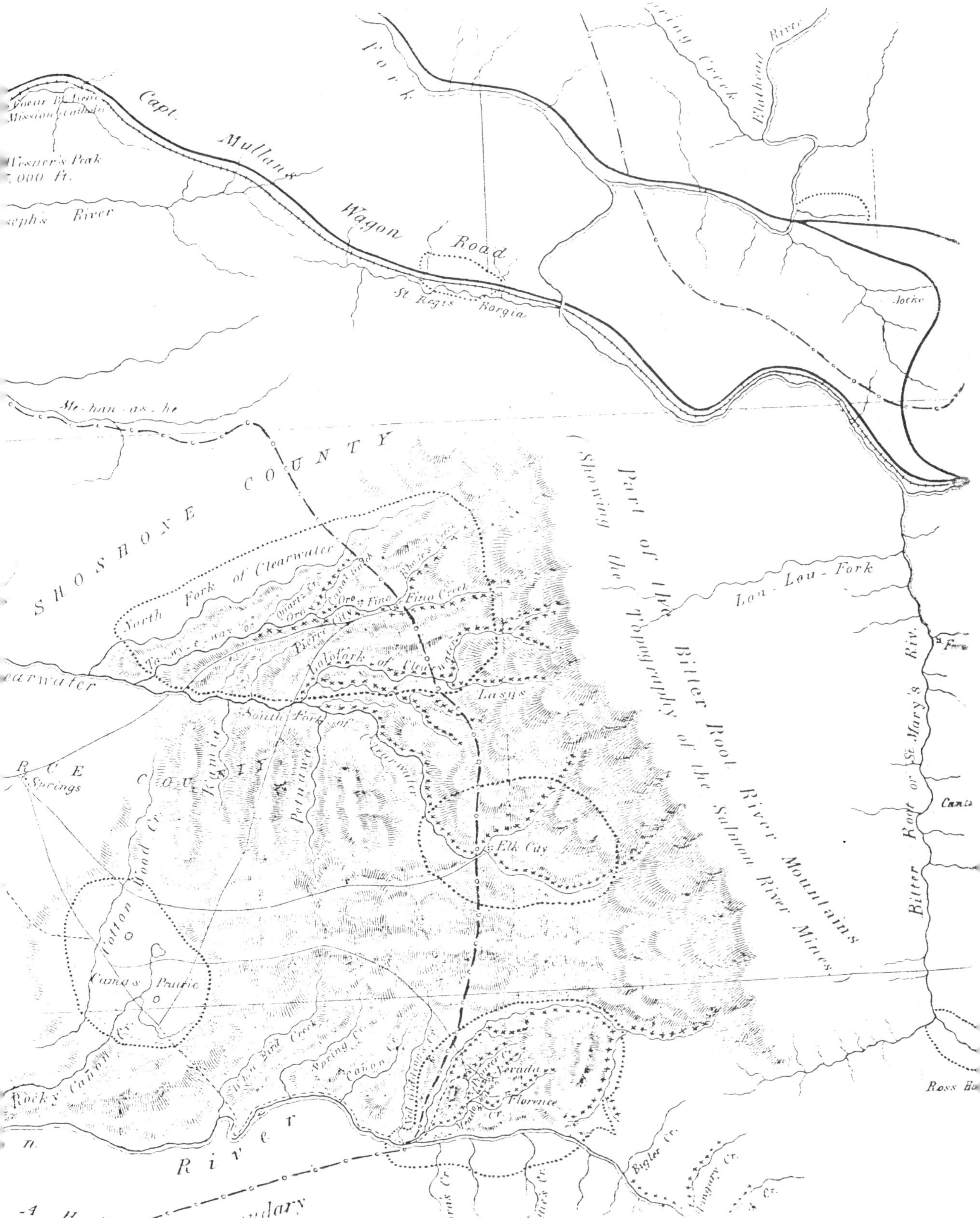
Capt. Mullans Wagon Road
Fork
Flathead River
Coeur D'Alene Mission
Wesner's Peak
7,000 Ft.
Joseph's River
St. Regis
Borgia
Jocko
Me-han-as-he
SHOSHONE COUNTY
North Fork of Clearwater
Oro Fino
Fino Creek
Pierce City
Lolo fork of Clearwater
Clearwater
South Fork of Clearwater
Lou-Lou-Fork
Part of the Bitter Root River Mountains
Showing the Topography of the Salmon River Mines
Bitter Root or St. Mary's Riv.
Elk City
Springs
Cotton Wood Cr.
Camas Prairie
Rocky Cañon Cr.
White Bird Creek
Spring Cr.
Cañon Cr.
Nevada
Florence
Bigler Cr.
River
Boundary
Ross Ho

his name would be stricken off their list. When Bill demurred the man struck him in the face, told him "to carry himself straight," and the incident was over.

I packed into the Kootenay mines in British Columbia in the summer of 1866 and 1867 with merchandise from Walla Walla. I also packed into Helena, Montana, again—two trips during the same year. I first heard the name Roosevelt there in Helena in the 60s. Then there used to be a camp out at Oro Fino gulch—above Helena—known as Roosevelt City after a relative of Theodore Roosevelt, an uncle I believe, who had come out from the East during the gold rush and who became interested in owning ground and operating a ditch in Oro Fino gulch. What his first name was I don't recall. They later changed the name of that camp to Unionville. Lots of gold went out of the Helena camp in those days. Along in August, 1866, a million and a half of gold left Helena in one shipment made under strong guard from Diamond City on Confederate gulch. This was taken out to Fort Benton, and then shipped by boat down the Missouri River.

I also made trips into the Boise Basin when Bannock City was burned there in 1865 and 1867. The two fires were exactly two years apart. On account of there being another Bannock City east of the Mountains in Montana they later changed the name of the Idaho Bannock City to Idaho City. The Montana Bannock City was over on Grasshopper creek, a tributary of the Beaverhead. The town sprang up there when they made the gold discoveries in 1862. From the time I went into these mines in 1861 up to the spring of 1863 all these mines of Oro Fino, Pierce City, Salmon River, Boise Basin, Bannock Idaho, Bannock Montana, Virginia City, Helena were in Washington Territory, and the territorial population had suddenly expanded from 11,000 or 12,000 to over 60,000 people. Probably 25,000 of these were in what is now Idaho and a like number in what is now Montana.

We had no post offices, or courts at first and the great distance from Olympia and the loss of time and the difficulty and expense in attempting to transact official business, and the utter inability of officials in distant Olympia to keep track of these transient mining camps, or to enforce laws soon led to a demand for the creation of Idaho Territory. The Idaho Territory bill was passed by Congress and signed by Lincoln in March, 1863, and the Territorial Government was organized a few weeks later. They hadn't much more than organized Idaho when the same conditions required the creation of another territory east of the mountains. Montana Territory was created the next year, 1864. This town of Bannock City was the temporary capital of Montana Territory and the first legislature met there in the winter of 1864. Virginia City then won the permanent capital; Helena didn't get it until ten years later. In this way, during the time I was packing, I've seen the same and identical mining camp in three different Territories.

The main pack trail or road from Walla Walla and Umatilla into the Boise Basin followed the general route of the old emigrant road across the Blue Mountains and by the Grand Rounde and Powder Rivers to Burnt River and down that stream to the Snake River at Olds Ferry near Farewell Bend. Crossing the Snake River we continued up the Snake to the Payette, and up the stream to what was known as Horseshoe Bend; then on up the Payette to Porter Creek or Jackass Gulch, and by trail up Porter Creek into Placerville. Most of the early travel and freight into the Basin went this way. An improved road later left the Payette this side of Horseshoe Bend and went up Shaefer's Creek into Placerville. These roads used to be lined with pack trains, saddle trains, stage coaches and individual parties on foot or horseback with their pack animals going into or returning from the mines.

Part of the road went over alkali flats and the dust stirred up by one party didn't get a chance to settle before the next party came along. A lot of road houses sprang up along the route to supply food and liquor to the travelers. I recall some of them. There was one at this Horseshoe Bend; another at Payette's ranch; others at Burns', Thompson's and Shafer's ranches. Meals were regularly a dollar a throw and whiskey two bits or more a swallow.

Bannock was the largest town in the whole Boise Basin, and the destruction of most of the supplies in the fires resulted in the immediate movement of the pack trains to that point to meet the needs of the thousands of miners then in the Boise Basin. Fully 2,000 mules loaded with freight went out of Walla Walla alone in the next year with supplies for the Boise mines. The freight charges from Walla Walla were then 35 cents to 40 cents a pound. It took about two weeks to reach Boise Basin and during the summer months, from 1863 to 1867 or later, there were about a hundred pack trains and freight trains operating from Wallula and Walla Walla into the towns of Boise Basin; ten to twelve such trains used to arrive and unload every day during the summer. The mountain roads were practically impossible in winter and early spring. Pack animals proved to be superior to ox teams, and horses and mules seemed to be less troubled by the alkali dust along the road.

This old mule train drawing is from Harper's Weekly, 1869.

Mule Train Packing

In addition to pack-trains, there were "saddle trains," and express riders ran along this route. The owners of saddle trains provided passengers with saddle horses and equipment, and furnished supplies for the trip, and permitted each passenger to carry a small amount of personal baggage, about 25 pounds. Tracy & Co. of Portland, and Rockfellow & Co. ran expresses for a time from Portland and Walla Walla into the Basin. Wells Fargo & Co. later bought them out. George F. Thomas who had a stage line from Walla Walla to Lewiston joined hands with J. L. Ruckle, built a toll road and established a stage line from Umatilla, over the Blue Mountains, into the Boise Basin, and Henry Greathouse put on a saddle train to connect with the Oregon Steam Navigation's boats at Umatilla,—a saddle train was a horseback passenger line. This was a common means of transportation in those days. Later on a stage coach was brought up from California and Greathouse's saddle train was changed to stage coach line. About this time Hill Beechy, who had sold out his Lewiston Hotel and livery stable to the Dwight Brothers, moved to the Boise Basin and put in a stage line making direct overland connections from California into the Boise Basin, by way of Star City, Nevada. Hostile Indians stole his stock, attacked his stages and put him out of business.

There were a number of good sized boom mining towns in the Basin in those days. Bannock, or as later called, Idaho City, was the chief one. It was at the junction of Moore and Elk Creeks, about 35 miles northeast of Boise City and had a population then around 6000 people. Buena Vista, or Buna Vista Bar was on the opposite side of the creek. Smith was Wells Fargo & Co's., agents there; Alf Eoff was the agent for Holliday's Overland Express. Greathouse & Kelly, and Ed. Pinkham had the local express lines to the neighboring camps—Centerville, Placerville, Pioneer City and Rocky Bar. I would say that there was fully 5000 people in these towns. Pioneer City was first known by the refined name "Hog-Em." Brittan, Ralston and Ridge & Co. were agents in charge of the Wells Fargo & Co. express at these places. Boise City itself was then a town of 2000 to 2500 people. I presume the way "Hog-em," or as later called Pioneer City got its first name was from the greedy character of the first locators who wanted to "hog it all," or stake and hold all the good ground to the exclusion of later comers.

Captain John Mullan, he of the Mullan road fame, was also a prominent stage line operator in and about these Boise Basin camps. As I recall he promoted an overland stage line from Boise south to Chico, California. This line like Hill Beachy's was also greatly bothered by Indian attacks. Captain Mullan was also at one time interested in one of the stage lines running from Umatilla Landing into the Boise Basin.

I was in Bannock, or Idaho City, in 1865 when Fred Patterson shot and killed former Sheriff Pinkham. Patterson was a self important fellow, he boarded at the City Hotel, and didn't get up until about eleven or twelve o'clock in the morning when it was his habit to go to the barber shop, just above the hotel and get shaved before getting his noon breakfast. Patterson was a man who had a superior air about him and didn't mix much, he belonged to secessionists of the camp. He had some kind of a grudge with Pinkham who was a union man. I was camping down at Dry Creek then and had come into town that morning and was on the street at the time.

Well, that morning when Patterson reached the barber shop he found Pinkham in the barber shop getted shaved; Patterson just walked over to the barber chair, drew out a big dragoon six shooter and placing it against Pinkhams ear shot him through the head. After he was shot Pinkham jumped up and ran to the front door, and fell there dead. I was right there at the time and saw him lying there in a pool of blood. Patterson "gave himself up," and his crowd being in control of things he was later acquitted by a packed or intimidated jury. Patterson was a bad man, he had killed a sea captain in Portland, and murdered a number of men in "self defense." When a bad man wanted to commit murder with impunity he picked a quarrel with his victim and killed "in self defense."

In going from Walla Walla to the Kootenay mining district we traveled over the Mullan Road to the crossing of the Touchet River, the site of the present town of Prescott; thence to the Snake River, which we crossed sometimes at Silcott's or Lyons Ferry and sometimes at Texas Ferry. We struck the Mullan Road again at Rock Creek and followed it to the crossing of the Spokane River, or Herrin's Bridge, as the place was then called. This bridge was located near the Idaho line, about a half mile above the place where Col. Wright had, in 1858, coralled and slaughtered several hundred head of horses belonging to the Indians of this region. The bone piles of "Horse Slaughter Camp" were then very noticeable in those days. From this point the Kootenay trail diverged from the Mullan road and we proceeded northeast to the present site of Rathdrum, Idaho, which in 1866 was called Conner's Ranch; thence to a ferry on Pend Oreille River, about 12 miles below the present town of Sandpoint. This ferry was called Sinna-acquateen, then the unsettled county seat of the unorganized Kootenai County of Idaho Territory. From the ferry the trail led east along the north side of the river to the site of Sandpoint.

I was in Walla Walla when the *Mary Moody* was built and launched on Pend O'Reille Lake. In going into the Kootenai mines in the spring of 1866 with the George Dacres pack train we crossed at Sinnaccateen ferry and went up the river to the Sandy point; there we boarded the *Mary Moody* and transported mules and packs to the mouth of Pack River. The boat would easily hold 60 mules and their loads. From Pack River the trail led to Stampede Lake, about 15 miles from Pack River; thence to Bonner's Ferry on the Kootenai River. The trail then led down the Kootenai River for 15 miles to a camping place and thence across to Moyie River, a distance of about 16 miles; thence up the north bank of the Moyie a distance of 40 miles, when it crossed this river at Peavine Prairie, near the home of Ogden Howell, a trapper. This was at Moyie whence the Moyie River takes its rise. It was on this trail that I met the Camel pack train. From this crossing the trail led to St. Joseph's Prairie, where the Canadian revenue officers were located; thence to the upper crossing of the Kootenai; thence five miles to Kootenai town, on Stud Horse, or Wild Horse, Creek. The name Kootenai is spelled with an i in the United Stated and a y in British Columbia.

Kootenay was a very rich strike and the placer mines there were worked for several years. Kootenay gold dust was very fine and was taken in at $18. an ounce. In 1866, there were perhaps a half-dozen stores and two or three saloons in Kootenay. The rates for carrying freight to Kootenay, either from Walla Walla or Wallula ranged from 40 cents to 60 cents a pound. We had to pay duty on all the goods that were transported across the international boundary. This was paid in the town of Kootenay.

I first saw the camel pack train on the Kootenay trail. I had heard of it but had never before seen it, and as we were camping near by some of the boys said; "Let's go down and see it;" so we went down. This was on the Moyie River in 1866 or 1867; this camel train was packing into the Kootenai mines from Walla Walla or from Fort Hope. Our own pack train was a pretty big one, 125 to 130 mules, and there was great excitement on account of the report that the camels were frightening and driving the mule trains off the pack trails. There were three men in charge of these camels, one cook and two packers. The train was a small one; only about six animals. The camels kept up a constant bleating when being loaded and unloaded, and on hearing them or smelling them a mule train would at once stampede right off the trail. The camels were equipped with a pack saddle, instead of the aparejoes such as we used, and they carried 800 to 900 pounds apiece, as against 350 to 500 for our pack mule. The camels browsed on brush, and could get along anywhere a mule or cayuse could. On August 3, 1928, Mr. Lewis received a letter from Mr. Watt saying: "The only other camel pack trains I knew of were of little importance; one having about twelve camels, the other about half that number. They were loaded, as near as I can recall, from Umatilla or Wallula to Bannock City, in the Boise Basin. They only made one trip. On this trip they stampeded a large mule train, doing so much damage that after leaving Bannock City they were taken out to Salt Lake. These were the only two camal trains on the Boise trail. The British Columbia camel train, that I met at the Moyie river on the trail to the Wild Horse mines, was I believe, packing from Fort Hope at the time."

J. Normansall was the postmaster at Kootenay Post Office. Among the Kootenay merchants were:—George Dacres, of Walla Walla; Oppenheimer & Co., Brenner & Co., Buckley and a man named Manuel.

Wallula, on the Columbia River, was the center of a freight train and stage line transportation to Colville, and the Columbia River, and Kootenay placer mines; and also to the early mining camps along the Pend Oreille, and into Western Montana, along the route of the old Mullan road. A great deal of freight was landed here for distribution through Walla Walla and Lewiston. Alvin Flanders, of Flanders & Fenton was postmaster and Wells Frago & Co's local agent. Margaret Moon ran the leading restaurant, and Parks & Hall ran the Wallula Hotel and Hazard Stevens, son of Gov. I. I. Stevens, was agent for the Oregon Steam Navigation Company operating the boat line from Portland.

Walla Walla was quite a bustling little town, and a great stage and freighting center. Captain Ankeny and Sons had one of the leading stores. Condon was Wells Fargo & Co's agent; George Thomas & Co., operated the Walla Walla, Wallula and Umatilla stage line into Boise City. Thomas Tienney and J. F. Abbot had the principal livery stables.

During the freighting days I made one trip to Colville with a load of general merchandise for Marcus Oppenheimer, with the pack train of George Dacres of Walla Walla—Dacres was later the owner and proprietor of the Dacres Hotel of that town. There was a military wagon road from Walla

Walla to the U.S. army post, Fort Colville; as a pack trail it was fairly good, but the ground was so soft that wagons were not run over the route until late in the summer. We set out for Colville in the early spring of 1866—in March—and midway encountered a heavy snow storm that raised the waters of Deep Creek and other streams and made the ground so soft that we had difficulty in crossing with our loaded animals.

Oppenheimer's store was on the flat which became known as Marcus Flat—the site of the old British Boundary barracks. Oppenheimer had evidently taken possession of the abandoned buildings after the British Boundary Commissioners left, and he had his store in one of the large log buildings. The Hudson's Bay Company post, Fort Colville, was but a short distance south of Oppenheimer's. The trail ran close to it, and I visited the place and met the Trader, Angus McDonald, while I was there.

Pickney City, officially known as the Fort Colville Post Office, was then the only town and trading center for northeastern Washington and the Pend Oreille country, and the Kootenai and Big Bend, or Columbia River mines of British Columbia. Park Winans was postmaster at Pinckney City, and among the leading merchants were; Louis Abraham & Co., Daniel H. Ferguson & Co., and Edward T. Smith. John Shaw and Stephen M. Harris ran saloons, and John Hoffstetter operated a brewery. A Mr. White was captain of the *Forty-nine,* and W. W. Briggs, purser of the steamer.

I packed into the Warren's Mining Camp in the years 1868 and 1869 with freight from Lewiston. Warrens then had 1100 or 1200 people and there was a great deal of mining yet going on there in the camp. The trail to Warrens went through the old Florence diggings and that district was still a pretty lively mining camp in 1868 and 1869; but the population had dwindled to 600 or 700. S. S. Fenn was postmaster there. I quit packing into the mines in 1869. My last packing trip to the mines was from Umatilla to the Sumpter Camp in the Blue Mountains in Oregon. After that I was engaged for a time packing to the government forts in Eastern Oregon. When they located Camp Klamath, Camp Warner, and Camp Harney I packed out to these posts, and I was with the command that went after Captain Jack and the Modocs in 1868. I was a civilian employee, and not enlisted in the regular service, and hence am not technically an Indian war veteran, though I've had all the experiences and have felt all the thrills of a real Indian fighter.

In my day I handled lots of gold dust and I became somewhat of an expert in judging it. Gold from the different mining camps nearly always had different assay values. The current trade value of this gold was always somewhat under the mint or assay value. The gold was classified and called "light," "heavy," "coarse" and "fine." The Florence gold dust was worth about $12 to $14 an ounce—it was "light" gold; we took it in at $12. Oro Fino gold dust was worth about $16 an ounce—trade value $15.00. Kootenay gold coined at $20. to $22. to the ounce, and was taken in trade at $18. Some of the Elk City gold was of nearly equal but the average was lower and we took it at $16. Bannock gold was $15 to $16 an ounce. We took in Boise gold at $15. The Warrens gold was light was taken like that of Florence at $12.00.

I knew the men operating Tracy & Company's Express into the Oro Fino mines in 1861. This was later consolidated with Wells Fargo & Co. I also knew Mossman, of Mossman & Co's Express operating between Oro Fino and Walla Walla. Most of the miners carried their own gold dust and nuggets out of the country. The stores of the camp accumulated large amounts of gold dust as there was no banking business to speak of. Most transactions were in cash, and the usual medium of exchange was gold dust. Every store, saloon and other place of business had its own gold scales. The man taking in the gold usually weighed it, and one had to be pretty well acquainted with the localities the gold came from in order to know the correct value to place upon the gold dust. Usually, at least in the first year or two, miners were pretty honest and would correctly tell you where their gold came from. Anyone frequently handling gold dust soon acquainted himself with the peculiar color and texture of the dust dug out in the various mining sections, so that he could tell, almost at a glance, where the gold dust offered to you came from, and then its corresponding correct trade value.

Folks now-a-days haven't much conception of the richness and extent of those early placer mines. Why, the whole country from the Blue Mountains to the eastern slope of the Rocky Mountains, and from southern Idaho far north into British Columbia was just one big gold field. There was rarely a stream that wouldn't "pan at least a color," and practically every square mile of that vast territory was some time or other traveled over and prospected by some of those prospecting parties in the latter 50s and early 60s. I traveled a good part of it more than once with pack trains. True, most of it would just pan a color and wouldn't pay to work but the rich diggings were numerous and bushels of gold were actually taken out of claims in camps like Florence, Boise and Helena, and gold dust and nuggets were for a time so plentiful that they were weighed and handled by the pound, and stored in tin cans and other handy receptacles.

This gold only found its way gradually into the assay offices and the mints at San Francisco and Philadephia. A large amount of it was for years used as a circulating medium throughout the Northwest. Nearly everyone then used to carry around some gold dust in his "poke." As a general thing the returning miners packed out their own gold. In those days it was not an uncommon thing for a miner to come out from the mines carryng anywhere from $2,000 to $3,000 or more in gold dust and nuggets on his person and concealed in his pack. I don't think over half in gold dust then mined was shipped out through the express companies, or made a matter of record anywhere.

The rush into these placer mines brought along a lot of men from every walk of life. Many of them especially old California, Oregon and Australian gold miners, were solely bent on searching for and mining gold. Lots of others went to the mines to carry on every day business pursuits catering to the miners' needs; others came simply to work at the high wages there paid. Along with these hordes of people,—I say hordes because there were actually thousands and thousands of them. Why in camps like the Boise Basin it was like as if there was an army camped there—little groups and bodies of men, scattered over the flats, gulches and hills, as far as you could see. At night one could see the camp fires everywhere.

Well, as I started to say, along with the hordes came the unscrupulous gambler, the bandit, road agent and murderer, and the usual camp following of dance hall girls, sporting house women, and other citizens of the underworld that preyed upon both the miner, the merchant and the legitimate laborer. At first we had no courts, or any means of law enforcement, and early in the discovery of placer fold in the Clearwater, the Salmon River, and the Boise and Upper Missouri River water sheds, this latter law-breaking element reaped a rich harvest. For a time in the absence of any organized peace force the honest men in camp hesitated to take any individual stand against them, and incur their ill will; so the criminal element carried on their operations in each camp for a short time with apparent impunity.

It was somewhat difficult in those days for law abiding men to know just whom to trust among their neighbors and acquaintances in camp. In the end the situation in nearly every camp became so bad that the better element finally had to band themselves together in a secret body to enforce law and order. These organizations were named "Vigilantes;" I never knew why they called themselves by that Spanish term. I suppose that the earlier organizations in California first got the name there. These bodies of courageous citizens soon helped to make the mountain trails and pathways safe for the honest miners, merchants and packers. A few hangings in a community were usually sufficient to check the graver crimes of murder and robbery. Life in those days was lightly valued by some, and the deeds of these road agents and of the vigilantes for a few years in the early 60s beat beyond comparision any dime novel or present day fiction of the wild and wooly west I've ever read.

I was intimately acquainted with most of the people connected with the Magruder murders. I knew Hill Beachy well and was often a guest at the Luna House. Magruder was a packer and trader with headquarters at Lewiston; he was married to a girl named Arthur, a sister of Sam Arthur, one of the early hotel men at Spokane Falls, Washington Territory. Magruder's wife lived in McMinnville, Oregon and as a young fellow I knew both her and her sisters. Magruder himself I knew and had often associated with as a fellow packer. He packed out of Lewiston. People about Hill Beechy's livery stable at Lewiston knew Lloyd Magruder's mules and equipment and this was the first clue of the murders. Dan Dwight later long in the merchantile business at Lewiston associated with his brother Henry Dwight, was one of the guards over the murderers in bringing them up from Portland. It was freezing weather when they were brought into Lewiston. Bill Page, who stood "State's evidence" and made a full written confession of the crime was released and afterwards did chores about Beechy's Hotel at Lewiston.

In October, 1863, Magruder, who had been on a packing trip to the Beaverhead mines, sold out his goods and part of his pack train, at Virginia City, Montana, and was returning to Lewiston with considerable bullion. It was said he had between $16,000 and $20,000 in gold dust. Enroute back, he and some companions, Charles Allen, William Phillips, and two brothers, Horace and Robert Chalmers, were brutally murdered in the Bitterroot Mountains in the most cold blooded manner by men who had accompanied Magruder from Lewiston in the guise of packers for the sole purpose of robbing him. The murderers were "Doc" Howard, Christ Lowry, J. P. Romain and Bill Page. Their capture and trial is a part of the history of those days.

Mule Train Packing

These murderers had hung around The Dalles and Lewiston for some time before their departure with the Magruder party in August, 1863, and I knew them. Daniel Howard was a well built, good appearing man who had evidently studied medicine at some time and he was generally known as "Doc." James P. Romain was an idler and gambler, and acted as cook on the fatal trip. Christopher Lowry was a man from down in Oregon, of rough and reckless disposition. He was a blacksmith by trade and had been with Capt. John Mullan on his road building expedition a few years before. William Page was a worthless, no-account fellow, easily led. He came from over in the Klickitat country, in Washington Territory, opposite The Dalles.

Hill Beechy used to say that he had dreams of the murder of his friend, even before the murderers appeared in Lewiston. There was some suspicion of them at that time and they were questioned some, but there being then no suspicion or evidence of a murder or other grave crime, they were permitted to go and soon made their way down to San Francisco. One of the murderers, Christopher Lowry, had been a schoolmate of Mrs. Magruder's down in Oregon. After his arrest she went to the jail and asked him what he and his companions had done with Magruder. He afterwards told his jailors that he didn't ever want to see her again. This man Lowry, himself, had killed Magruder with an axe.

Bill Page, who had turned "State's Evidence," was killed a couple years later at Lewiston, Idaho, by Albert Igo. Al was a touch, worthless character himself, laying up with sporting women, and of a very bad reputaton. Page at the time was reduced to the job of carrying water from the river up to the house of ill fame where Al Igo was then staying. Igo had been drunk the night before and in a quarrel had been beaten up by Bill Page. He was in an angry mood the next morning and when Page entered the house carrying a pail of water in each hand, Igo blew his head off with a shot from a shot gun. This was on December 25, 1866. Al Igo immediately fled from Lewiston but he was followed by his brother Bill Igo, who overtook him and persuaded him to go back with him and stand trial. Everyone in Lewiston thought the killing of Bill Page was a good act, and after some pretense at an investigation, Al Igo was turned loose. Many years ago a man named Chapman wrote up the history of the Magruder murders. "Doc" Howard, Lowry and Romain were executed at Lewiston, Idaho, on March 4, 1864, after a short trial.

James W. Watt

This Magruder murder trial was the first case we ever tried in the Courts of the new Territory of Idaho. It was tried before Judge Samuel C. Parks, Judge of the Boise County, or Second Judicial District, who held a special term of Court at Lewiston for that purpose. The defendants were found guilty on January 20, 1864, and sentenced to death. A detail of soldiers from the 4th U.S. Infantry at Ft. Lapwai formed a guard around the gallows at the hanging on March 4, 1864. Hill Beechy's expenses at running down and bringing back the murderers were something over $6,000 and later the Idaho Territorial Legislature passed a special appropriation bill to reimburse him. Hill Beechy later sold out his hotel and livery stable in Lewiston to the Dwight Brothers and moved to the Boise Basin where he operated stage lines to Nevada and California. He died down in San Francisco about fifty years ago.

My cousin, D. M. Jesse, was once held up by road agents on this side of the Clearwater River, about 25 miles out from Oro Fino on a return trip to Lewiston. Jesse was on the alert, and when he met the holdups he shot at them and chased them into the brush. Pack trains were frequently held up on the Boise trail.

I knew Alex Carter well. He was one of the cargadores of Bledsoe and Creighton's pack trains, running into Florence in 1861, and I went into Florence with him on that first trip in 1861. He was a tall, powerful, fine appearing man, and was not afraid of anything. He always bore a fine reputation among us packers. He packed on the Boise road until late in the fall of 1863. Then he came into Umatilla, collected his wages and procuring a couple of mules, started out for Montana. A few weeks later, on January 16, 1864, way up at Hell Gate, Montana, some 15 or 20 miles from Missoula, he was seized and after a hearing at Higging's store he was taken out to the corral and hanged by the Montana vigilantes. They claimed Alec to be a member of the notorious band of road agents headed by Henry Plummer, who was sheriff at Bannock and Virginia City, Montana in 1863.

We, who knew Carter, couldn't believe that he was guilty. He was a man from a good family back in the east, and was a trusted employee of Bledsoe and Creighton, often having thousands of dollars of their money and property in his hands. He was always recognized among us as a law-abiding man. I remember an occasion at Florence in 1861 when Mat Bledsoe, a nephew of his employer, and a worthless gambler, got on a drunken tear in his uncle's store at Florence. In those days, stores kept an open whiskey barrel in the back of the store with a tin dipper, and customers buying a considerable bill in cash, or settling up old accounts were customarily invited back and asked to "have a drink." Mat was a mean drunk, and he got out his gun and began 'shooting up' the store. His uncle wanted him to get out, but he refused to go. Alex Carter, who had been standing over by the fireplace, finally got up and says in a clear commanding voice, "Mat, put up your gun and get out of here," and when Mat hesitated he went right up to Mat, grabbed him by the coat collar, and kicked him out of the store.

I don't think Alex Carter deserved hanging. Later I went by the place where he was hanged and tried to locate his grave, but I couldn't. All the accounts I've heard or read state that Alex Carter continued to assert his innocence, even as he was being strung up. He's written up in the Vigilantes of Montana.

Along with Alex the vigilantes hanged Johnnie Cooper. Cooper was from down in the Willamette Valley, Oregon, and had also borne a good reputation until he got mixed up with bad associates in Montana. His folks lived in the East. The old California and Oregon gold miners were as a whole a very steady, law-abiding lot of men who were steadily working for their "stake." Most of the trouble and lawlessness of those times was caused by the shiftless, moneyless adventurers about camp who were too lazy and shiftless to do hard work. So much gold dust was carried about that it was a great temptation to road agents.

Mule packing of freight will soon be "one of the lost arts." The settlement of the country and the building of railroads and wagon roads, put us out of business nearly fifty years ago. I haven't seen an aparejo for twenty years. Most of our packing terms were Spanish, picked up by the Forty-niners from the Mexicans in California. What is an Aparejo? Why it is Spanish pack saddle made of leather and stuffed with moss, dry hay or grass—anything handy; it protected the mule's back from any rubbing of the load, and equalized the weight of the pack on the animals. They were far superior to the American pack saddle.

With each aparejo there was a "caronie," a fancy Spanish embroidered pack blanket, which was laid over the "sweat" blanket; on top of this went the "bed" blanket, of best wool and costing in those days from $15.00 to $20.00. Each "caronie" was embroderied with a distinctive design or flower figure, and each mule had its own "caronie" and pack. In camp the packs were arranged in a hollow square or rude circle if possible—if not they were strung in a row, two packs high, with the aparejos on each side, the lash rope lying under the

James W. Watt

A HALT TO TIGHTEN THE PACKS.

A POOL IN THE DESERT.

loads. The aparejos were well made and often ornamented and cost from $35.00 to $60.00 apiece.

Arriving in camp, after unloading one of the packers rode the bell mare out into the hills for pasture, the unloaded mules followed along behind. We usually had a white mare for bell mare, and the mules would rarely ever leave her vicinity. Sometimes a strange mule had to be watched for a few days until he got acquainted with and accustomed to the bell mare. On the road we'd get up early, before day light; while the other men were breaking camp and preparing breakfast, one of the packers would go out into the hills and locate the bell mare. Finding her, he would ring her bell and the mules from all around would answer by bawling out, and starting towards the bell mare. The packer would then walk around to see whether all the mules were in sight and coming in; when satisfied of this he would mount the mare and ride into camp. If he started off on a trot or run, the mules would follow at a trot or run.

On the pack trail we would roll out of our blankets at two or three o'clock in the morning, and while some were bringing in the mules the rest would break camp and the cook would get breakfast.

When the animals came in the bell mare was brought to the head of the packs, and the men began to holler at the mules "Get in there," and the mules would herd in around the packs. Then the men quickly placed the "hackamores" or headstalls on the mules, and tied them together, all facing in towards the rigging which was usually arranged in a hollow square or circle if the ground permitted.

The mules secured, the men sat down to their breakfast. This finished, they hastened to load the animals, while the cook cleaned up and packed his outfit. Each mule had the same aparejo and load during the entire trip, unless it was overloaded or disabled. In the first days of my packing we averaged only 15 mules to the man, later this was increased to 18 ot 20 mules, and at last it was customary to give two men 36 or 40 mules to handle. It was some work, I'll tell you. The average time was 1½ to 2 minutes for two experienced men to pack a mule. Four men would load 40 to 60 packs in 45 to 50 minutes. The men who worked in pairs were called the 'near' and the "off" packer. With a big train, the pairs would work near together, and while his partner was finishing the tie the odd man would act as a "booster" assisting the other two men with the loading of their pack, by helping to lift and place the heavy load. Experienced men would pack a hundred mules like clockwork. The "sweat" blanket was first put on the animal's back—carefully smoothed out so that there were no folds to rub and gall the animal's back during the day; then the "caronie" and the bed blankets were added; then the aparejo; then the load. If it was 400 pounds of flour each packer would first lift his 200 pounds of the pack up on to his side of the animal; then throw the "swing" rope and "swing the packs;" then the lash rope was thrown on top and lashed in the "diamond hitch" and the animal was ready for the trail. The "cinch" of the aparejos was 8 or 10 inches wide; it was cinched tight about the protesting mule by two men. The "ladigo" strap or cinch strap was 2 inches wide, and it was well greased so as to slip easily in tightening.

Packers were always armed. The cook on the bell mare usually carried a shot gun or rifle; the packers were armed with rifles and revolvers. We carried "cantinas" or leather fittings with two pockets with flaps fitting over the saddle horn for holdng little conveniences. I usually carried a revolver in the right side of my "cantina." Some carried their gold dust there. It was also a handy place for a flask of whiskey.

As quick as the last load was securely packed and tied, the cook mounted the bell mare and rode off; the loaded mules of the train followed behind without command. The cook usually carried a water bucket. The packers mounted their horses and took convenient places along the train as the loaded mules went along in single file. The "Boss" packer rode behind. The start was usually made by six o'clock in the morning. The train moved at a fairly rapid walk to the next camping place. These averaged about 15 miles apart, at convenient places where there was feed and water. The next camping spot was reached, according to the distance, sometime between ten o'clock and noon, when camp was made for the day.

If it stormed, or there was a heavy rain, the train remained in camp for the day; if it rained or stormed on the road, then the train kept steadily on until the camping place was reached. Packs were protected from the weather by "mantos;" large 9 by 12 pieces of canvas which were laid over the packs in camp, or folded over the loads on the trail in heavy rains or snow storms.

Part of the art of packing consisted in understanding how to put up the freight for packing; this was done by the packers. The loads were proportioned to the various animals in the train according to their strength; a mule would carry from 300 to 500 pounds, an occasional animal could carry considerable more. Flour was packed in burlap sacks of heavy weave holding from three or four sacks of flour aggregating a weight of 150 to 200 pounds. Liquor was carried in kegs; usually in an "8 cask" holding 28 gallons. One large barrel made an average mule load. The most difficult load I ever packed was some solid steel shafting, each piece 8 or 9 feet long and weighing about 200 pounds apiece.

In those early days when there were no roads, only just pack trails, we packers would pack anything—quartz mills, burial caskets, cans of powder, or even a piano. One time Bob Grostein's train packing a quartz mill from Lewiston into the Warrens diggings had a single piece of steel casting—a crescent shaped plate on which the stamps hammered—weighing 667 pounds. It was carried the entire 100 miles by one husky mule, over one of the most difficult trails I've ever packed over. In going to Warrens diggings we dropped down Slate Creek from Florence into the Salmon River canyon; we crossed the river by a wire cable suspension toll bridge built in 1864-5, and about 10 miles below Florence by trail. This bridge was just wide enough for a loaded pack mule to pass through. The toll was 75 cents per loaded animal and 35 cents a head when we returned. Going into and out of the canyon the trail led along the edge of precipices with perpendicular cliffs above and below. Mules occasionally lost their footing and fell. I recall that one mule, loaded with women's apparel for the demimondes of the camp, fell off this trail and went crashing down through the tops of the pine trees below. In the fall the pack broke and all manner of women's apparel was scattered over the tree tops to flap and flutter in the breeze to the amusement of we packers and everyone passing along the trail. The poor mule was of course killed.

Many packers were owners of stores, or traders, buying their own cargoes and packing the merchandise into the mining camps to sell on their own account. Arriving in camp they would wait for Sunday and then, unpacking their goods would often sell their stock out in the day. If there was a remainder it was disposed of to some merchant at a price which covered the packing charges and a fair profit on the goods.

I was acquainted with most of the master packers, and I worked at one time or another for most of them. D. M. Jesse and H. H. Snow were in partnership as D. M. Jesse & Co. They had a large pack train, Jesse had charge of it, while Snow handled the merchandise business. When the packing business fell off in 1866-7 they sold out their train and Mr. Jesse went into the stock business at Walla Walla; while Mr. Snow went into business at Lafayette, Oregon. His son Wilbur is in the bank at Dayton, Washington.

Old Andrew Lafevre was a partner with "French Louie" in a train of 65-70 mules at Walla Walla operating into Helena and on the Boise road in

1864-1865. George Dacres of Walla Walla had a train of from 45-50 mules. I went with this train to Helena in the fall of 1865; Charlie White, his "boss packer," who lived at the mouth of Dry Creek, left him at Helena and I took his place. I was then twenty-two years old. I handled the train for two years. Frank and Mat Lowden at the mouth of Dry Creek, near Walla Walla, also owned a train of about 125 mules, and packed into the various camps. They were likeable and popular men, and Frank was later elected County Commissioner of Walla Walla county.

Jones and Dalton ran a pack train of 125 packs out of Lewiston. Bludsoe and Creighton also ran a train of 65 to 70 packs out of that place. D. M. Jesse and John Thompson had a train of 40 packs running out of Lewiston and Pierce City. Bob Grostein also ran a train of 100 mules out from Lewiston. Virgo Little was boss and packmaster of a train of 40 or 50 mule team running from Umatilla into the mines. Bill Sperry and his brother also ran a train out of Walla Walla; later they sold out and went into the flour milling business at Pendleton, Oregon. Their brother-in-law, Manse Chrip, located the first farm above what is now Pendleton; part of his original holdings are in the present townsite.

Most of the large packing outfits had their headquarters at Walla Walla, but they packed out of Wallula, Umatilla, Walla Walla or Lewiston, wherever they could get the freight. One trip might be into Florence or Warren or Mormon gulch over beyond Burnt river; the next might be to Helena or Virginia City; our next trip might be into the Kootenai or Columbia River mines, some 400 miles Northeast across the international boundary line into the Province of British Columbia, or south into the Boise Basin. In my 10 years on the trail I packed into every good sized mining camp in the Northwest.

Lize Dove and his brother had a train of about 25 mules running from Lewiston into Elk City and Florence. I worked with this train in 1864 and made one trip into Florence. Lize died here in Spokane about five or six years ago, while his brother Tom died down in the Palouse country years ago. At one time they were large land holders there and had wealth, but they went busted in the hard times of 1894.

In packing days I worked with Johnnie O'Hearn's pack trains from Walla Walla; also with George William's train from The Dalles, Oregon, packing into the military posts. O'Hearn had the contract for Camp Hearny supplies.

Mule Train Packing

The highest freight we ever received was around a dollar and a quarter a pound, received when we went into Bannock, Idaho after the fire; the lowest freight received from Umatilla into the Boise Basin was twenty cents a pound for the three hundred mile pack. Pack trains used to be of all sizes, from a small train of but five or six animals up to large trains numbering a hundred or more animals. It all depended on the amount of business. Some pack train proprietors had several hundred pack animals on hand and kept a number of pack trains in constant and regular operation on a fairly regular schedule from outfitting points into the various mining camps. An average pack train into the larger mining camps would average say about twenty-five pack mules. Trains of twenty-five pack animals were easier to handle than larger trains on account of the greater convenience in making camp and finding feed. The size of the trains was increased or reduced, or additional pack trains placed on the route, as the immediate traffic needs of a mining camp required.

Packing was a trade which called for both skill and strength. In loading and unloading we usually worked in pairs; one on each side of an animal. Each man had to swing his half of the heavy pack up from the ground on to the animal's back, and hold it there on one side of the pack saddle with one arm, despite the kicks and bites and protesting lunges of the mule, until he and his partner had it securely lashed there by means of a "diamond" or other hitch. To round up, saddle and load and unload twenty-five such packs a day, and to meet all the emergencies of the trail was a real man's sized job. If a mule got down, you'd have to get him up; sometimes you'd have to unload; other times the mule would succeed in unloading for you. Some mules got real mean and tricky, and would try to rub their packs off against trees; or would lie down and try to roll them loose. It took a good tie to stand such treatment, even if you rode up at once and started the animal along with the train again. If straps and ropes broke you had to splice them; you had to mend the pack saddles; sometimes you had to shoe the mules; sometimes animals got sick and you had to nurse them. Worst of all sometimes the packs broke, or sprung a leak, and you had to devise means—way out along in the wilderness—to save your cargo.

And talk about roads and trails; principally there weren't any; leastwise a tenderfoot would have gotten lost a dozen times a day just trying to follow some of them. We penetrated into the most remote and inaccessible places; over all kinds of country and in all kinds of weather conditions. At the head of the trains, long before you met them you could hear the tinkle of the bell on the bell mare. For convenience in located stray animals, we sometimes use to hobble new pack horses and mules at night, unless a coral was handy to put them in. We sometimes placed a bell on animals that were giving to straying away for convenience in locating them when turned out. On all ordinary occasions the "bell mare" kept the pack animals together.

Heavy snows in the mountains usually forced us to quite our packing ventures along in November; then when we returned from our last trip we drove the pack animals down to winter at some ranch in one of the lower, warmer valleys along the Columbia River where there was plenty of good feed and water.

As I've said this packing business called for both skill and strength; the lifting of heavy weights—sometimes a barrel of whiskey—right up from the ground to the pack saddle required a young muscular man with a stout back and good arms and shoulders. Quite a few of our packers were Mexicans, brought up from California, and employed on account of their special experience and skill in such work. In addition to taking care of the freight and animals a man had to prepare camp and cook his own meals, and act as is own valet, and wash and mend his own clothes. We often camped out in rain and snow and severe cold without even a tent. With the larger pack trains one man was usually taken along as cook for the outfit.

Besides all this work we had to be on the alert to perserve our own scalps. On the trail there was always more or less danger from attacks by hostile Indians, and murderous road agents. If they didn't kill you they might run off your horses and mules, or rob you of your freight. It wasn't an easy life by any means. At first there weren't even any trails, and we often had to strike out over virgin country and find our own way, with no one to enquire of, in case we got lost. When we came to bad places we had to work our train around them; cutting a way through fallen timber; working around or across a bad rock slide, or detouring impassible cliffs. Why, often there weren't any bridges, or ferries; when we came to streams we sought for a ford; if there wasn't any, we swam our animals across, and if there were no boats available, we then built a raft and rafted our outfit and goods across. Sometimes when a spring freshet had made a stream impassible, we had to sit down and wait until the high water passed. The whole packing business called for courage, hardihood, and endurance; that wasn't all, you had to have skill, daring initiative, gumption, and loyalty.

Mule Train Packing

In many respects the packers I worked among were, take them all in all, a rough, lawless and profane bunch of men; but they were brave, hardy and extremely loyal and trustworthy towards their employers.

The few men in a pack train were often intrusted with train equipment worth ten to fifteen thousand dollars and with the delivery of merchandise worth as much or more again. Freights were paid in cash, and goods sold for cash, and the packers sometimes returned with as much as fifteen or twenty thousand dollars worth of gold dust, belonging to their employers.

In those early days of Washington, Idaho and Montana Territories there wasn't any money in the Territorial Treasuries to build needed roads and bridges, so the Legislature met the situation by granting individual franchises to build private toll roads and toll ferries and toll bridges. On a long pack like that from Wallula into the Kootenai country mines, or from Wallula into the Montana mines where there were many ferries or toll bridges to cross the toll charges for the trip might run up to five or ten dollars an animal for trip. All of this was added to the expense of packing ~~and~~ included in the freight charged.

As the trails were improved into roads, the toll bridges and toll ferries were installed the freight pack trains of the packers and muleteers were gradually replaced by wagon freight trains driven by "bullwhackers" and "Mule skinners." On anything like a passable road it was far cheaper to haul merchandise, than to pack it; bigger loads could be carried, better time made, and the expense of equipment and labor was greatly reduced.

Quitting the packing business, I remained in Oregon for several years before moving to Washington Territory and making my permanent home there. In 1883 at the time of the completion of the Northern Pacific Railroad, I settled in the Lance Hills district, nine miles southwest of Cheney, in Spokane County, then Washington Territory, and I have lived there ever since. My brother, Alex Watt, an old gold miner of the 60's also settled in that neighborhood.

James W. Watt.

James W. Watt

Spokesman-Review February 20, 1944

PIONEER DIES, 100 YEARS OLD

CHENEY, Wash., Feb. 19.—James W. Watt, old-time packer who helped develop the far west by bringing in supplies overland, died yesterday at the age of 100 at his ranch home 10 miles southwest of Cheney, where he had lived since 1884. While he never did packing with camels, he often told of seeing such work done in the old days in the southwest.

Mr. Watt was born in Salem, Ohio, on May 31, 1843, and came west with his parents to Oregon in 1849. He retired from farming about 20 years ago, but was still active enough to ride horseback over his ranch when well past 90, according to his son Arthur, with whom he had been living. Also surviving is his wife Lillie E., at the home, and a grandson, Everett, a teacher at Winlock, Wash., and a granddaughter, Mrs. Howard Franz, Seattle.

Funeral services have been set tentatively for 1:30 Tuesday afternoon at the Jerue funeral home in Cheney, with burial in the Lancehill cemetery.

Index

Y

COLOPHON

The James W. Watt, JOURNAL OF MULE TRAIN PACKING IN EASTERN WASHINGTON IN THE 1860's *was printed in the workshop of Glen Adams, which is located in the quiet country village of Fairfield, southeast Spokane County in Washington state. Fairfield is located on state highway 27 that runs between Opportunity and Tekoa. The James Watt work was first printed by Ye Galleon Press in 1978, and all copies were hard cased. This edition sold out some time ago, so a decision was made to print it again in paper covers. The 1978 typesetting was done by Robin and Bobi Pearson using a Compugraphic model 48 photosetter. Running heads are in 18 point Bologna, handset type. Text was set in 12 point Baskerville with a 15 percent enlargement of the pages. Title page and colophon page were designed by Pat Nigh and Susan Paulson, who also stripped the film and made the printing plates. Printing was done by Vern Stevens using a Model KORS printing press. Folding was done by Garry Adams using a 26 x 40 Baum Dial-O-Matic folder. Binding of paper covers was done by Garry Adams and Glen Adams. This was a fun project. We had no special difficulty with the work. Ye Galleon books carry a cargo of words...as long ago in the sixteenth and seventeenth centuries the clumsy wooden Spanish galleons carried rich cargoes from the New World to the Old.*